ATOMIC STRUCTURE

FOR KIDS

JOLPIC KIDZ

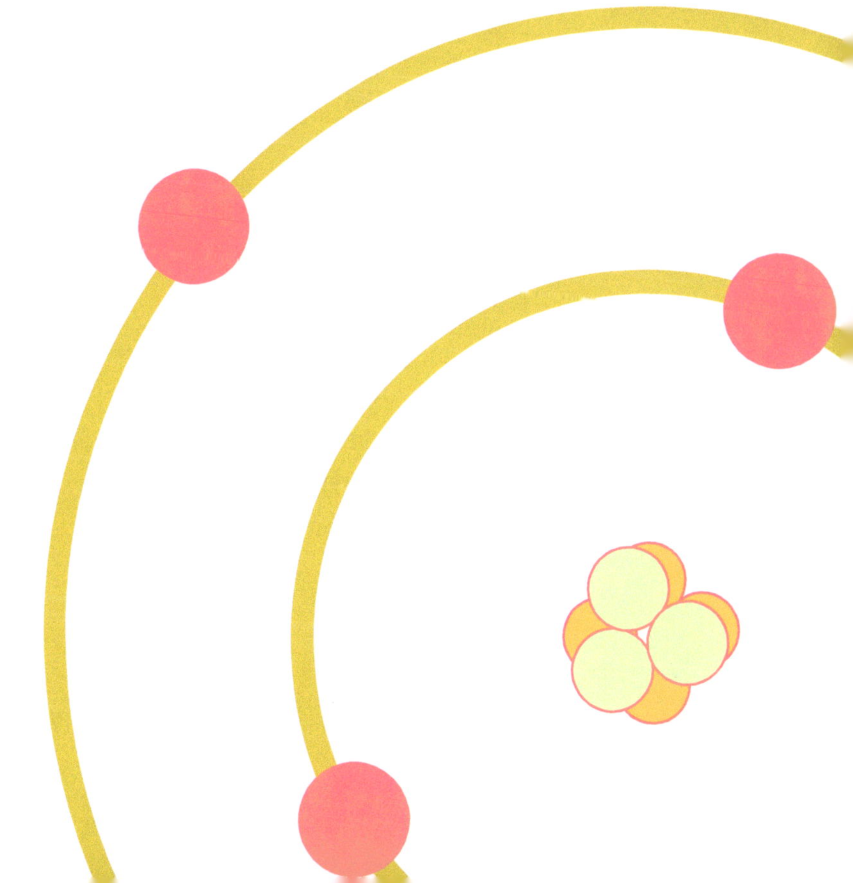

Atomic Structure for Kids

First Published: March 2022

Jolpic Kidz is a publishing company of educational books written for kids.

For more info, mail us at
jolpic@gmail.com

Copyright © Jolpic Kidz 2022

Everything is made of **ATOMS**

And rest of the things that you can see, are made of ATOMS.
You are unable to see wind, even it is also made of atoms.

ATOMS are tiny particles, we cannot see them.

How many types of Atoms are there?

There are only 118 kinds of Atoms existing in the universe, still we can see numerous things around us.

How is this Possible?

It is possible because an atom can combine with another kind in numerous ways, and they form new things.

We call those new things, COMPOUNDS.

For example

Two atoms of hydrogen combine with one atom of oxygen, and form a new compound, called water that we drink every day.

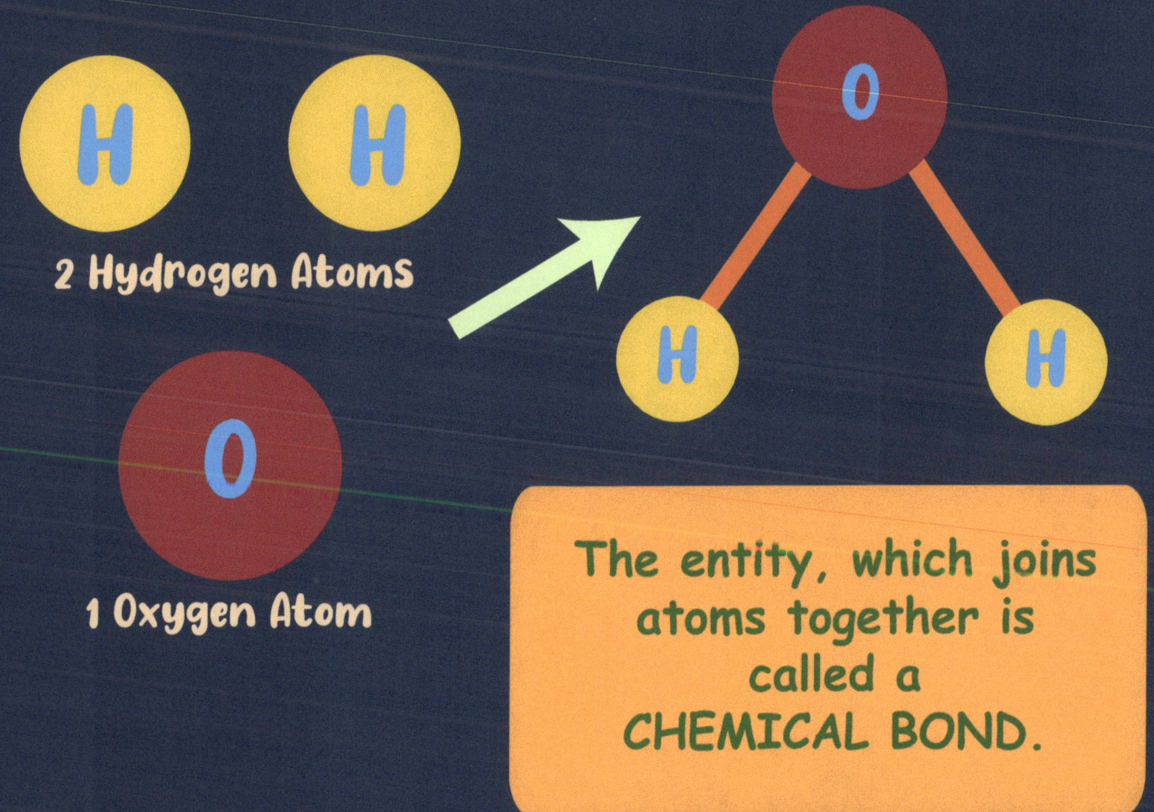

The combined form of atoms is called a MOLECULE.

How does an ATOM look?

Every atom is made of three fundamental particles.

They are,

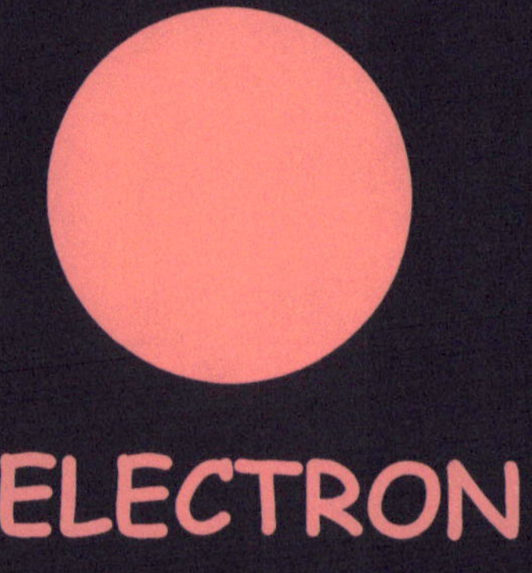 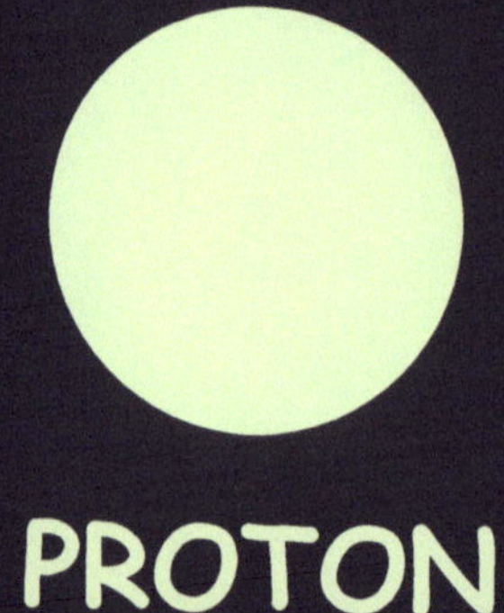 

ELECTRON   PROTON   NEUTRON

No one knows how these particle, ELECTRON, PROTON, and NEUTRON look like.
Consider, they look like balls.

Electron and Proton attract each other just like magnets.

ELECTRON

PROTON

They attract because they have charges.

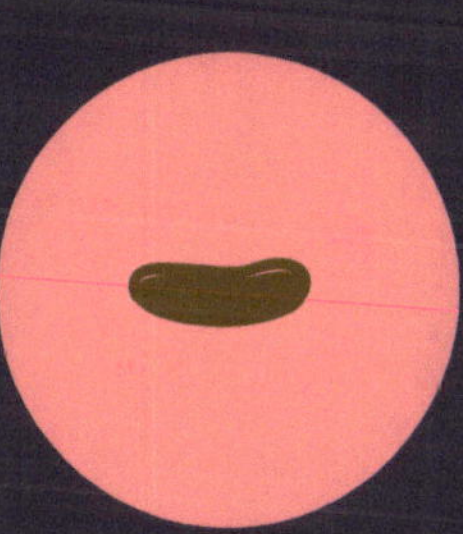

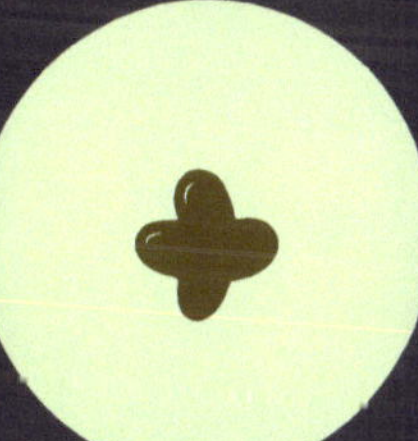

An electron has negative charge.

But a neutron has no charge

A proton has positive charge.

Atomic nucleus

Every atom has a nucleus at its center. It is the densest region of an atom formed by protons and neutrons sticking together. The mass or weight of any material originates due to the existence of atomic nucleus.

Since protons and neutrons possess more masses than electrons, elements having larger nuclei have greater masses.

Atomic Nucleus

Atomic Structure

In a simple model of an atom, we can say that electrons are revolving around the nucleus in circular orbits.

Atomic orbits are exceptionally larger than the size of the nucleus, and as a consequence, we can say that most part of an atom is empty.

Diagram of an atom

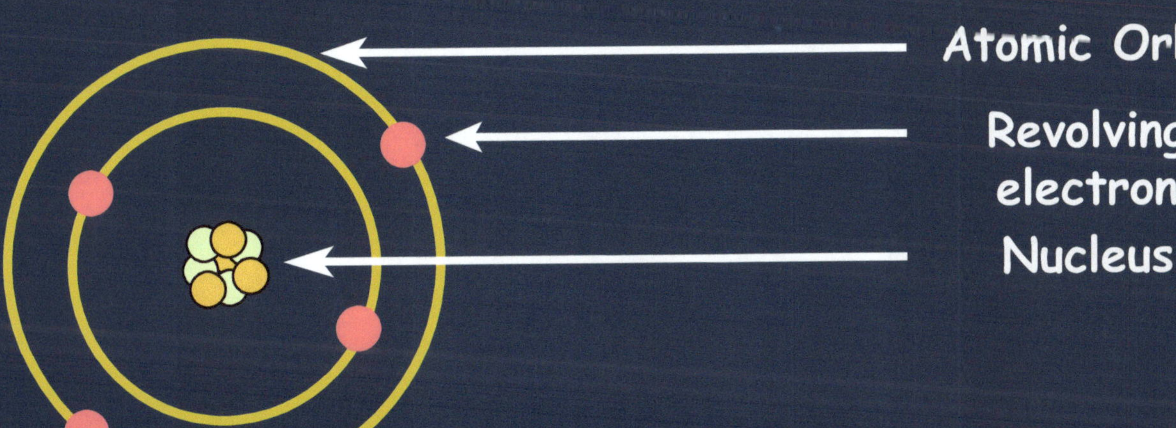

Now it is the time to learn about more!

Let's begin our discussion with hydrogen atom. We choose hydrogen atom because it is simplest among all other elements in the universe.

There is only one proton in the nucleus of hydrogen atom and an electron is revolving around it.

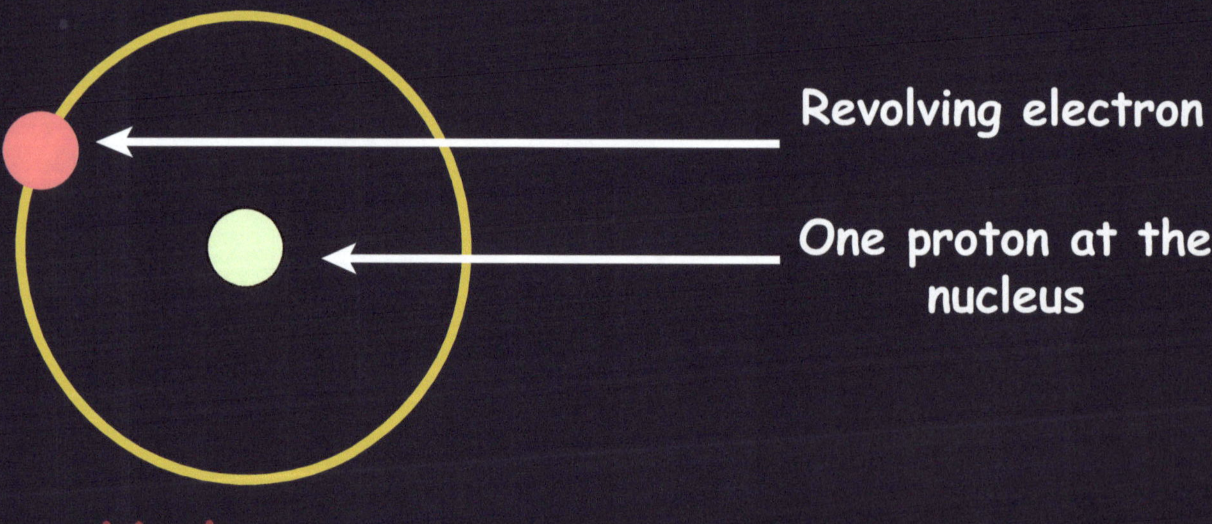

If it is true that a proton attracts an electron, why is the electrons not falling into the nucleus?

Electrons are not falling into the nucleus because of the existence of centrifugal force. This force balances the attraction between an electron and nucleus.

Centrifugal force is very common for any rotating object. For example, when someone rotates a stone tied to a thread over his head, the stone does not fall on the ground. Also, gravitational pull of the sun can not pull our earth into it due to the centrifugal force arising from earth's own rotation.

10

In spite of that explanation, according to the law of CLASSICAL PHYSICS, when a charged particle travels along a curved path, or a circular path, it loses its ENERGY, and falls into the NUCLEUS. Thus, an atom would have been destroyed.

But, it does not happen. An atom is quite stable.

All the laws of Classical Physics do not work in microscopic world. So, a new theory is introduced.

The new theory is known as,

QUANTUM MECHANICS

According to QUANTUM THEORY, electrons can travel along certain specific orbits. They cannot even stay in any region near the nucleus except those orbits.

While moving in a certain orbit, an electron does not lose its energy.

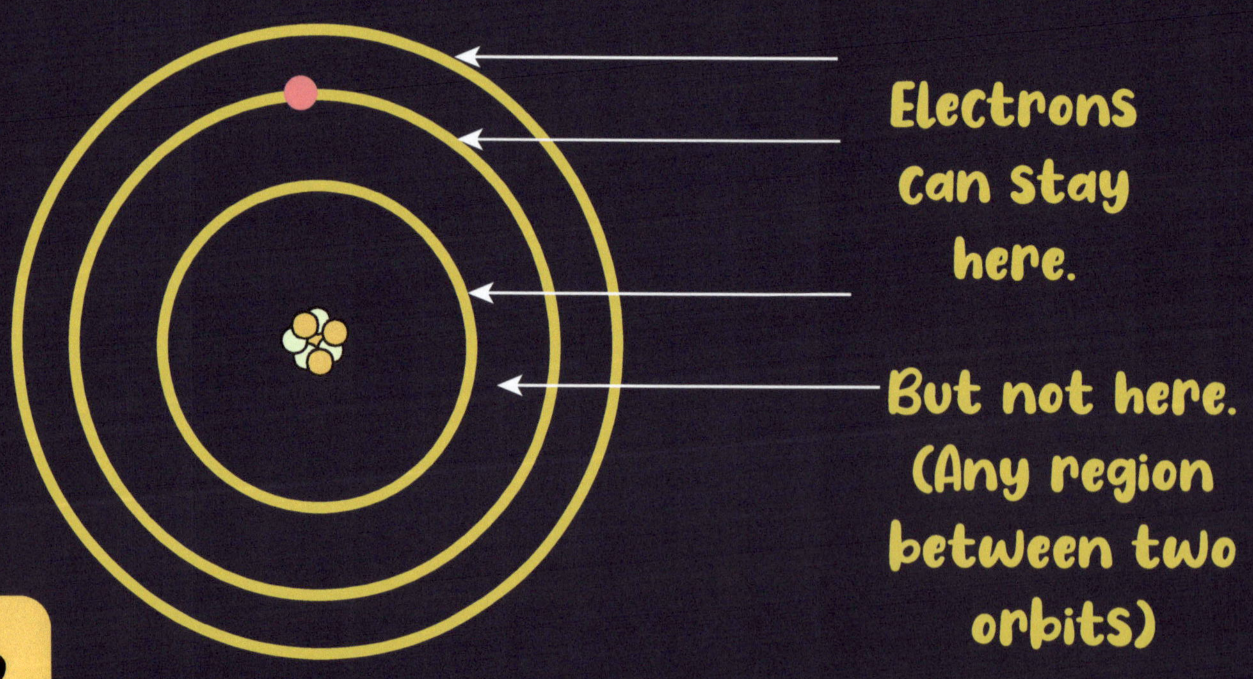

An electron can lose or gain its energy when it jumps from one orbit to another.

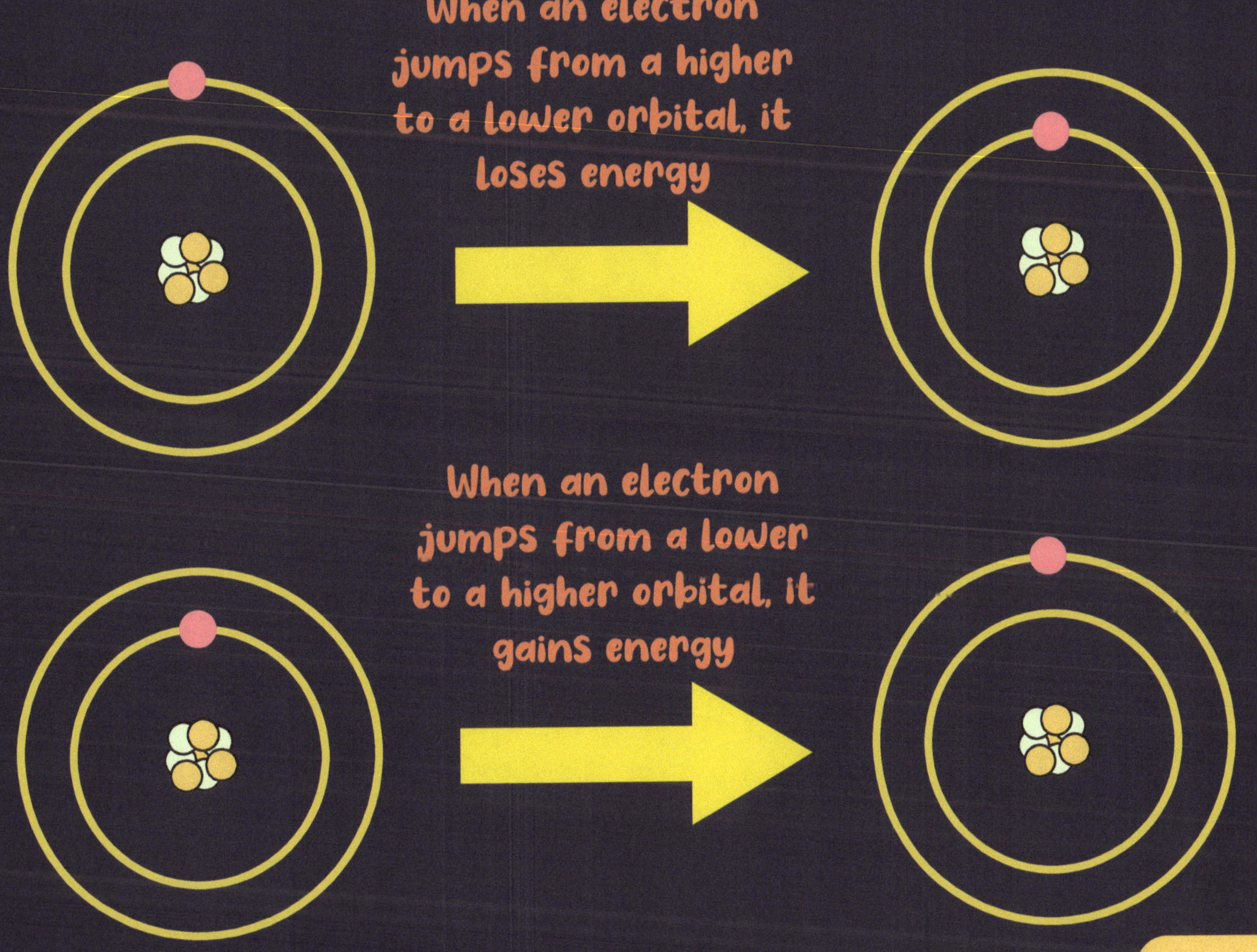

When an electron jumps from a
higher to a lower orbit, the electron
loses its energy in the form of
ELECTROMAGNETIC RADIATION.

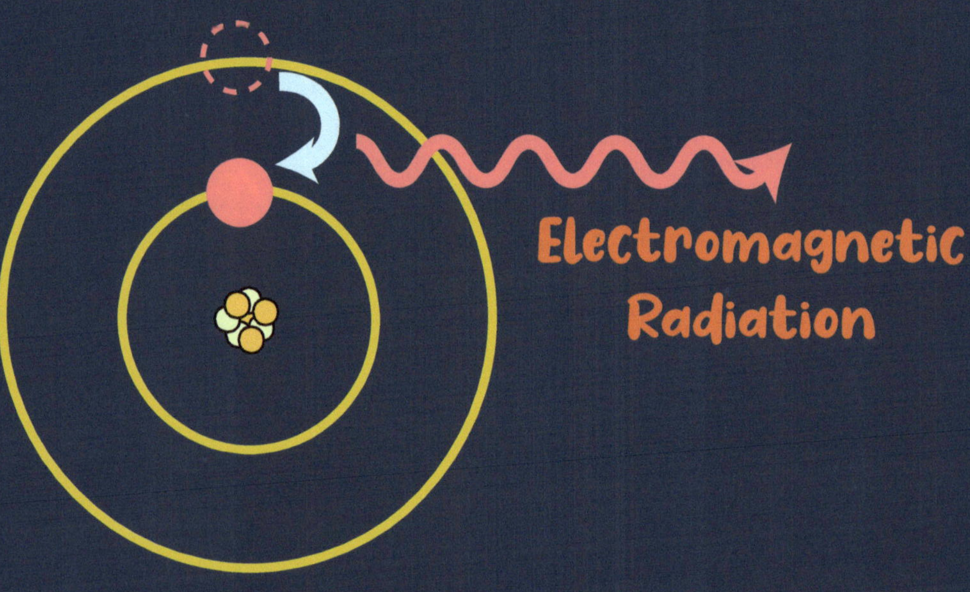

Electromagnetic radiation is nothing but
LIGHT. Those lights are may be visible in our
eyes, or maybe not. For example, all those
fire crackers show their colors due the
occurrence of such electron jumps.

For a particular atom, when number of electrons and protons are equal, the entire atom is charge-less, i.e. Neutral.

This is an example of a neutral hydrogen atom containing one electron one proton and zero neutron.

The charge an atom = Number of protons at the nucleus − Number of electrons in the orbits

If a neutral atom loses an electron, it becomes positively charged atom due the presence of one excess proton at the nucleus. On the other hand, if a neutral atom receives an electron, it becomes a negatively charged atom due to the presence of one excess electron in the orbit.

A charged atom is called an 'ion'. When an ion is positively charged, we call it CATION, and when it is negatively charged, we call it ANION.

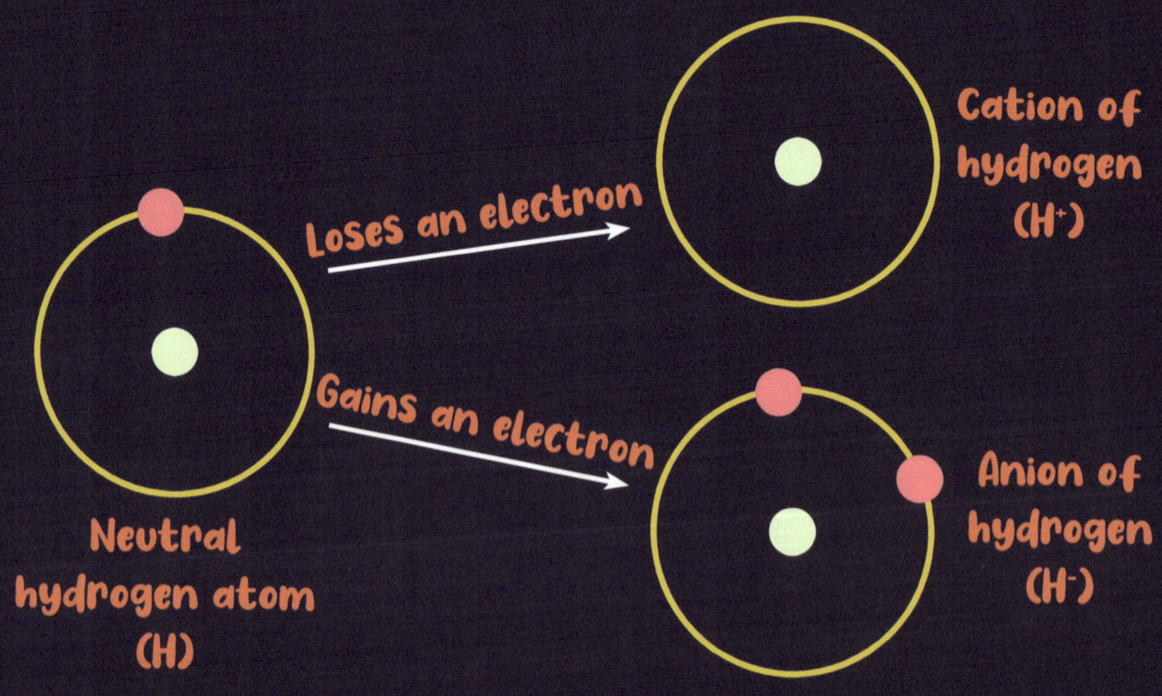

This is an example of Lithium cation, which contains three protons and two electrons.

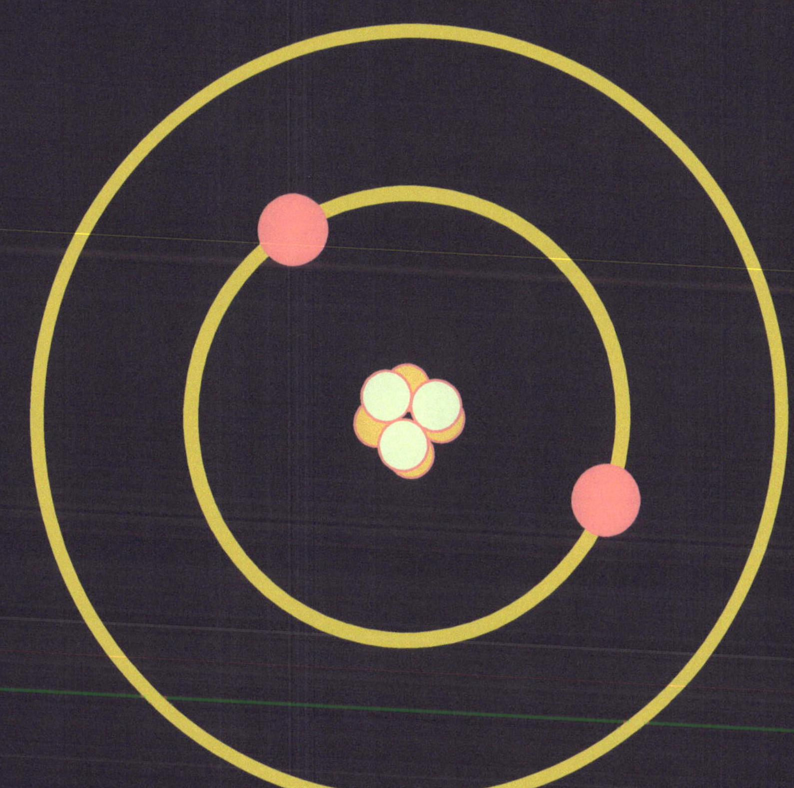

So its charge is:

3-2 = +1

The circular Orbits or Shells can be denoted by numbers according to their sequence.

Each numbering of shells is known as its PRINCIPAL QUANTUM NUMBER, and is expressed with letter 'n'.
Like, n = 1, n = 2, n = 3, n = 4.

Scientists had previously thought that the SHELLS were circular, but when they examined ELECTROMAGNETIC RADIATIONS of electron jumps, they considered SHELLS might be elliptical in shapes. Also, they found each SHELL is divided into SUB-SHELLS.

They named those SUB-SHELLS as,

s, p, d, f.

Those SUB-SHELLS are denoted by another QUANTUM NUMBER, known as, AZIMUTHAL QUANTUM NUMBER with letter 'l'.

AZIMUTHAL QUANTUM NUMBER of s SUB-SHELL is 0, i.e. $l = 0$.
AZIMUTHAL QUANTUM NUMBER of p SUB-SHELL is 1, i.e. $l = 1$.
AZIMUTHAL QUANTUM NUMBER of d SUB-SHELL is 2, i.e. $l = 2$.
AZIMUTHAL QUANTUM NUMBER of s SUB-SHELL is 3, i.e. $l = 3$.

The first SHELL with PRINCIPAL QUANTUM NUMBER 1 (n = 1) is not divided, and the SUB-SHELL for the first SHELL is 1s.

The second SHELL with PRINCIPAL QUANTUM NUMBER 2 (n = 2) is divided into two SUB-SHELLs, 2s and 2p.

The third SHELL with PRINCIPAL QUANTUM NUMBER 3 (n = 3) is divided into three SUB-SHELLs, 3s, 3p and 3d.

The fourth SHELL with PRINCIPAL QUANTUM NUMBER 4 (n = 4) is divided into four SUB-SHELLs, 4s, 4p, 4d and 4f.

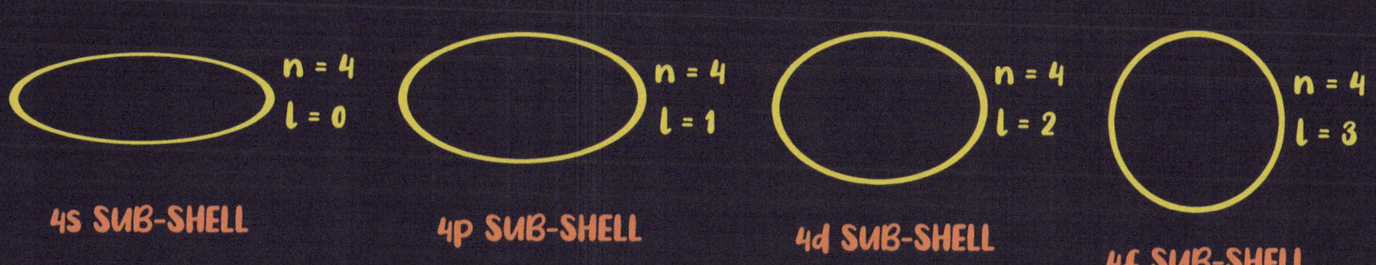

Magnetic Quantum Number

Scientists further discovered that those subshells further split when an atom is brought under a magnetic field (or near a magnet).
Those splitting forms are called, ORBITALS.

An s subshell does not split.

A p subshell splits into 3,

p_x, p_y and p_z.

A d subshell splits into 5,

d_{xy}, d_{yz}, d_{xz}, $d_{x^2-y^2}$, d_{z^2}

An f subshell splits into 7,

So, a new quantum number is introduced, called Magnetic Quantum Number.

For a particular subshell, Magnetic Quantum Number ranges from -l to +l including 0.

For example, the value of l of a p subshell is 1. Thus, the Magnetic Quantum Numbers are,

$$-1, \ 0, \ +1$$

Representing 3 Orbitals

Similarly, the Magnetic Quantum Numbers for a d (l = 2) subshell are,

$$-2, \ -1, \ 0, \ +1, \ +2$$

Representing 5 Orbitals

And, the Magnetic Quantum Numbers for an f (l = 3) subshell are,

$$-3, \ -2, \ -1, \ 0, \ +1, \ +2, \ +3$$

Representing 7 Orbitals

Magnetic Quantum Numbers are represented by the letter 'm'.

Electron Spin

Each orbital can hold up to two electrons (see next page for more). These two electrons must have different spins.

Another Quantum Number is introduced to denote the spins of electrons.

+1/2 and -1/2

Spin Quantum Number

In an atom, no two electrons can possess four Quantum numbers same.

If three Quantum Numbers, Principal, Azimuthal, and Magnetic Quantum Numbers are same for a pair of electrons, Spin Quantum Numbers must be different for them, +1/2 and -1/2.

This principle is known as Pauli's Exclusion Principle

As a consequence, it is an obvious fact that an orbital can not hold more than two electrons.

Let's verify the above statement with an example:

Consider for a while, a $3p_z$ orbital contains 3 electrons:

Corresponding Quantum Numbers:
For 1st electron: $n = 3$, $l = 1$, $m = 0$, $s = +1/2$
For 2nd electron: $n = 3$, $l = 1$, $m = 0$, $s = -1/2$
For 3rd electron: $n = 3$, $l = 1$, $m = 0$, $s = -1/2$

2nd and 3rd electrons have four Quantum Numbers same, which is impossible.

Thus a specific orbital can carry up to two electrons in up and down spin arrangements.

Shells	Sub-shells	Orbitals	Electrons' Spins	Number of Electrons
First	1s	1s	↑↓	2
Second	2s	2s	↑↓	2
	2p	$2p_x$	↑↓	2
		$2p_y$	↑↓	2
		$2p_z$	↑↓	2
Third	3s	3s	↑↓	2
	3p	$3p_x$	↑↓	2
		$3p_y$	↑↓	2
		$3p_z$	↑↓	2
	3d	$3d_{xy}$	↑↓	2
		$3d_{yz}$	↑↓	2
		$3d_{xz}$	↑↓	2
		$3d_{x^2-y^2}$	↑↓	2
		$3d_{z^2}$	↑↓	2

And so on....

Electrons possess different energies when they stay in different Subshells. Thus electrons should fill in lower energy Subshell first, then they occupy higher energy Subshells. This is known as Aufbau rule.

$$1s < 2s < 2p < 3s < 3p < 4s < 3d < 4p < 5s < 4d < 5p < 6s \ldots$$

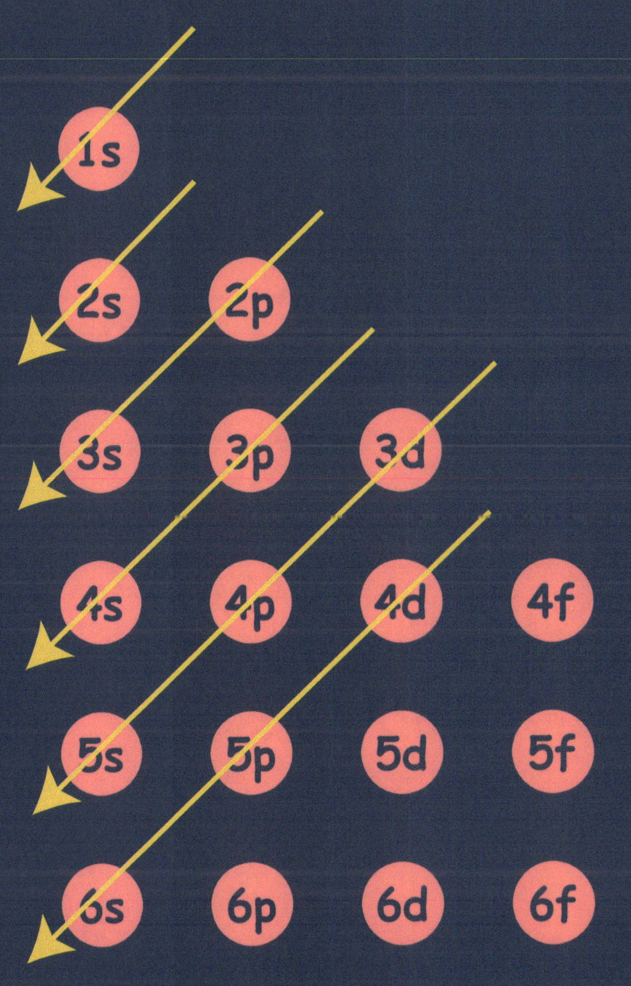

The order of filling up electrons in different Sub-shells is shown in this diagram.

Electron Configurations

Mass number:

The total number of neutrons and protons in the nucleus is called the Mass number of an atom.

Atomic Number:

The number of protons in the nucleus is called the Atomic number of an atom.

Examples

$$^{1}_{1}\text{H}$$ Hydrogen Atom

Hydrogen is the simplest atom, and its mass number is 1, and its atomic number is also 1.

So, number of proton at its nucleus is 1, and number of neutron is, $1-1 = 0$

A neutral Hydrogen atom should consist one electron, and since 1s orbital has lowest energy, the GROUND STATE electronic configuration is,

$$1S^1$$

Helium Atom

$${}^{4}_{2}\text{He}$$

The Mass Number and Atomic Number of Helium is 4 and 2 respectively.
Thus its nucleus contains 2 protons and (4 −2) = 2 neutrons.
A neutral Helium Atom requires 2 electrons to neutralize the protons' positive charges at the nucleus.

Thus the electron configuration is,

$$1S^2$$

Lithium Atom

$${}^{7}_{3}\text{Li}$$

Lithium contains 3 protons and 4 neutrons at the Nucleus.

So, the electronic configuration of lithium is,

$$1S^2 2S^1$$

Since the 1s orbital is already occupied with two electrons, the third electron should enter to the next higher energy orbital, 2s.

Beryllium Atom

$^{9}_{4}\text{Be}$

Mass Number: 9
Atomic Number: 4

Beryllium atom has 4 protons and (9-4 = 5) 5 neutrons at the nucleus.
A neutral Beryllium atom has 4 electrons. Thus the electronic configuration is,

$$1S^2 2S^2$$

Boron Atom

$^{11}_{5}\text{B}$

Mass Number: 11
Atomic Number: 5

Boron atom has 5 protons and (11-5 = 6) 6 neutrons at the nucleus.
A neutral Boron atom has 5 electrons. Thus the electronic configuration is,

$$1S^2 2S^2 2P^1$$

Carbon Atom

$${}^{12}_{\ 6}\text{C}$$

Mass Number: 12
Atomic Number: 6

A neutral Carbon atom has 6 electrons. Thus the electronic configuration is,

$$1S^2 2S^2 2P^2$$

1s	2s	2p

It is interesting to note here that among the three orbitals under 2p sub-shell, two are half filled instead of one full filled, because later is lesser stable than the previous.

Nitrogen Atom

$${}^{14}_{\ 7}\text{N}$$

Mass Number: 14
Atomic Number: 7

The electronic configuration of Nitrogen is,

$$1S^2 2S^2 2P^3$$

1s	2s	2p

Oxygen Atom

$^{16}_{8}O$

Mass Number: 16
Atomic Number: 8

The electronic configuration of Oxygen is,

$1S^2 2S^2 2P^4$

1s 2s 2p

2p sub-shell starts to fill up with opposite spin of electrons.

Fluorine Atom

$^{19}_{9}F$

Mass Number: 19
Atomic Number: 9

The electronic configuration of Fluorine is,

$1S^2 2S^2 2P^5$

1s 2s 2p

Neon Atom

$^{20}_{10}Ne$

Mass Number: 20
Atomic Number: 10

The electronic configuration of Neon is,

$1S^2 2S^2 2P^6$

1s 2s 2p

Now you are able to determine the electronic configurations for all the 118 elements. Although a few electronic configurations have exceptions, you will still be able to determine most of them accurately.

The Periodic Table of the Elements

Group→	1	2	3	4	5	6	7	8	9	10	11	12	13	14	15	16	17	18
↓Period																		
1	1 H																	2 He
2	3 Li	4 Be											5 B	6 C	7 N	8 O	9 F	10 Ne
3	11 Na	12 Mg											13 Al	14 Si	15 P	16 S	17 Cl	18 Ar
4	19 K	20 Ca	21 Sc	22 Ti	23 V	24 Cr	25 Mn	26 Fe	27 Co	28 Ni	29 Cu	30 Zn	31 Ga	32 Ge	33 As	34 Se	35 Br	36 Kr
5	37 Rb	38 Sr	39 Y	40 Zr	41 Nb	42 Mo	43 Tc	44 Ru	45 Rh	46 Pd	47 Ag	48 Cd	49 In	50 Sn	51 Sb	52 Te	53 I	54 Xe
6	55 Cs	56 Ba		72 Hf	73 Ta	74 W	75 Re	76 Os	77 Ir	78 Pt	79 Au	80 Hg	81 Tl	82 Pb	83 Bi	84 Po	85 At	86 Rn
7	87 Fr	88 Ra		104 Rf	105 Db	106 Sg	107 Bh	108 Hs	109 Mt	110 Ds	111 Rg	112 Cn	113 Nh	114 Fl	115 Mc	116 Lv	117 Ts	118 Og

Lanthanides	57 La	58 Ce	59 Pr	60 Nd	61 Pm	62 Sm	63 Eu	64 Gd	65 Tb	66 Dy	67 Ho	68 Er	69 Tm	70 Yb	71 Lu
Actinides	89 Ac	90 Th	91 Pa	92 U	93 Np	94 Pu	95 Am	96 Cm	97 Bk	98 Cf	99 Es	100 Fm	101 Md	102 No	103 Lr

Shapes of Orbitals

The behavior of electrons is very strange. They can be regarded as particles as well as waves, because in some experiments, an electron can behave like a particle, on the other hand, in some other experiments, an electron can behave like a wave. For that reason, electrons' positions are described with a mathematical term called PROBABILITY.

For example, when an electron is staying in an s orbital, the possibility of finding it is maximum in a spherical space around the nucleus.

When an electron resides in any of the three p orbitals, the probability of finding the electron in the space around nucleus looks like following.

Shape of an s orbital.

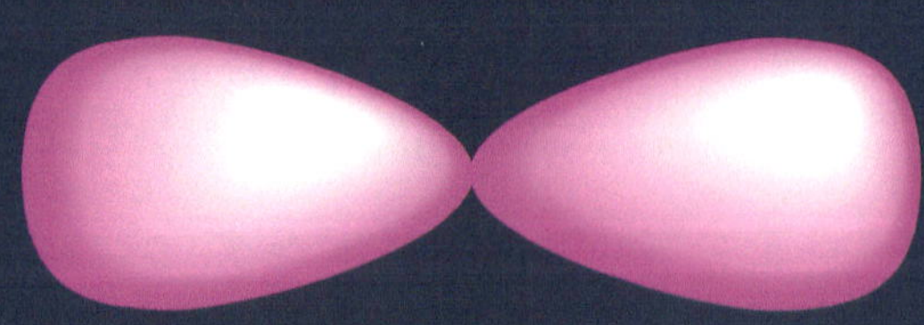

Shape of a p orbital.

The shapes of three p orbitals are similar; only their orientations will be different in three dimensional space.

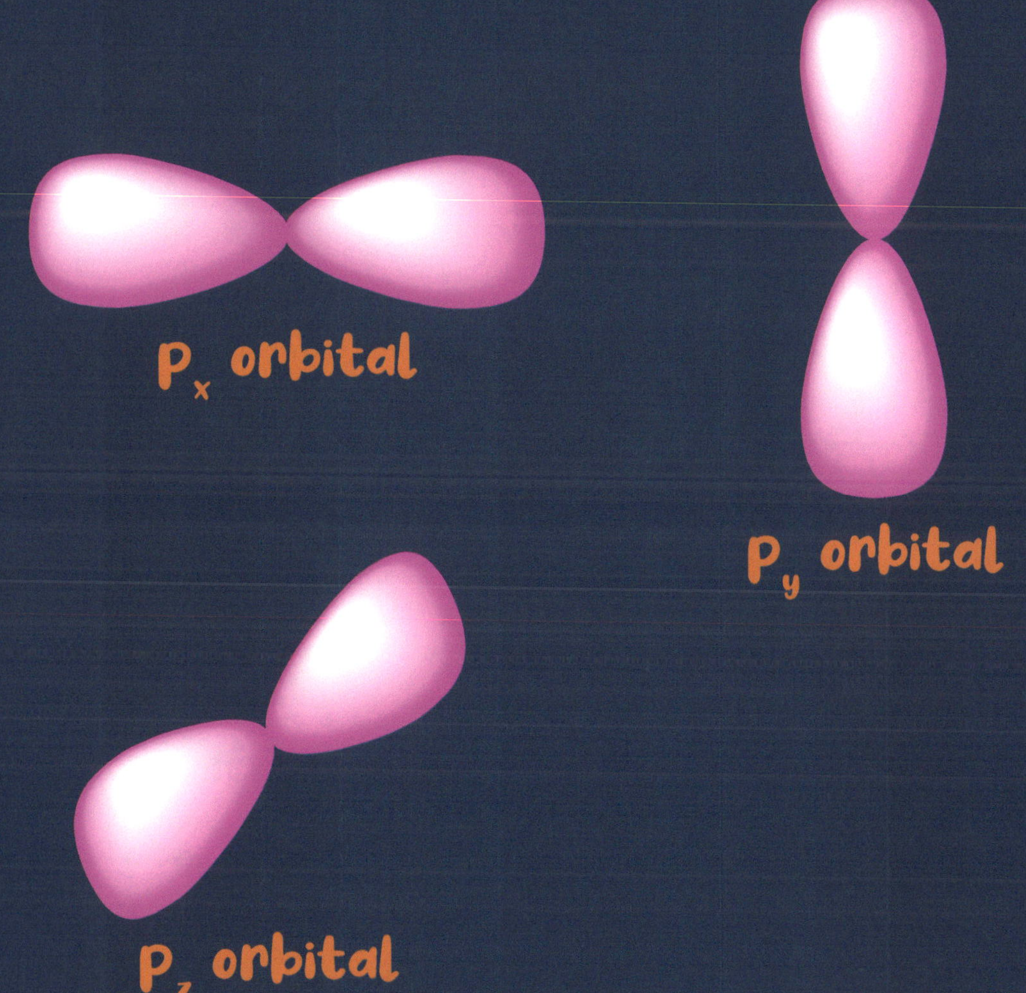

The End